Gorillas

by Helen Orme

ticktock

Copyright © ticktock Entertainment Ltd 2006
First published in Great Britain in 2006 by ticktock Media Ltd.,
Unit 2, Orchard Business Centre, North Farm Road,
Tunbridge Wells, Kent, TN2 3XF
ISBN 1 86007 963 6 pbk
Printed in Hong Kong
A CIP catalogue record for this book is available from the British Library.

We would like to thank our consultant Dr. Annette Lanjouw,
Director of the International Gorilla Conservation Programme

Picture credits
t=top, b=bottom, c=centre, l-left, r=right
Corbis: 15. Digital Vision: 4-5, 6-7, 8-9, 10-11, 12-13, 14, 18-19b, 22-23, 24-25, 26-27,
28-29, 30-31, 32. FLPA: OFC, 20-21. Nature Picture Library: 16-17, 18-19t,
Every effort has been made to trace the copyright holders, and we apologise in advance for any unintentional omissions.
We would be pleased to insert the appropriate acknowledgements in any subsequent edition of this publication.

CONTENTS

Words that appear **in bold** are explained in the glossary.

A VERY RARE ANIMAL

Mountain gorillas are one of the rarest animals in the world.

There are only about 700 mountain gorillas and they live in just two places in Africa. Both places are in **protected parks**, and are high up in the mountains.

The gorillas live in thick, misty **rainforests** on the mountain slopes. Some of the mountains are **dormant volcanoes**.

The Virunga volcanoes where about half of the mountain gorillas live.

YOUNG MOUNTAIN GORILLAS

A baby gorilla weighs about two kilograms when it is born.

For the first few months, the mother gorilla will hold the baby to her chest. As it gets older, she will carry the baby on her back.

When gorillas are about a year old they will climb trees and play on their own, but they never stray far from their family group.

GORILLA FAMILIES

A gorilla family will have one or two adult male gorillas, some babies and young gorillas, and some adult females.

Adult male gorillas are called silverbacks. This is because they have silvery fur growing on their back and hips.

Each gorilla family has a silverback as leader. The silverback protects the group.

SILVERBACK FACT

Silverbacks keep other animals away from their families by acting tough! They stand on their back legs and beat their chests.

LOOKING FOR FOOD

*Gorillas spend their day moving through the forest **foraging** for food. An adult gorilla can eat up to 30 kilograms of food every day.*

Mountain gorillas eat stinging nettles, bamboo, thistles and other leafy plants.

Although they are mainly plant eaters, sometimes they will look for worms, grubs and ants to eat.

BABY FOOD FACT

Baby gorillas drink their mothers' milk until they are about two years old.

Gorillas have strong jaws and large teeth for grinding up tough plant stalks.

DIFFICULT TIMES FOR MOUNTAIN GORILLAS

Mountain gorillas were only discovered by non-African people 100 years ago.

Many hunters came to Africa to kill them.

At last people realised that there were fewer and fewer gorillas.

The areas where the gorillas live became protected parks and hunting them was against the law.

MORE DANGERS FOR MOUNTAIN GORILLAS

Things slowly began to get better for mountain gorillas, but in 1990 a terrible war broke out in this part of Africa.

People tried to escape from the war by hiding in the mountains. Soldiers hid in the mountains, too. Some gorillas caught diseases from the people.

Many people live near the protected parks. The people need land for farming and wood for fuel. This destroys the gorillas' **habitat**.

People have to cut down the forest for firewood.

POACHERS

Poaching is another problem for the mountain gorillas.

Usually poachers are not trying to catch gorillas. The poachers are hunting for antelopes or bush pigs, which they can take home to eat. The poachers use **snares** to catch these animals.

Sometimes gorillas get caught in the snares that are meant for other animals.

In this picture **park rangers** are carrying the body of an adult gorilla that died when it was caught in a poacher's snare.

PEOPLE WHO HELP

Life is difficult for the mountain gorillas. But there are many people whose job it is to help them.

Park rangers patrol the forest. If they find an animal who is ill or in trouble, they can get help.

The mountain gorillas have their own special vets. The vets can go into the forest to treat the gorillas with drugs, or even do an operation.

In this picture, the rangers are checking on a gorilla who is ill. The silverback is trying to guard his family.

GORILLA TOURISTS

*It might seem a bad thing for **tourists** to visit the forests where mountain gorillas live. In fact, if this is done carefully, it can help the gorillas.*

Gorilla tourists bring money and jobs to this part of Africa. This means that protecting the gorillas is important to local people.

The tourists are only allowed to stay with the animals for an hour, and they cannot go closer than seven metres.

The mountain gorillas must be kept safe from human diseases.

GENTLE GIANTS

People often think gorillas are fierce. In fact, they are very intelligent and gentle animals.

HELPING THE MOUNTAIN GORILLAS TO SURVIVE

Mountain gorillas have survived hunting, war and disease.

As more and more people live closer to the gorillas, there is more chance of the gorillas catching human diseases.

More and more of the gorillas' habitat will also be needed for farms.

Helping them to survive in the future will not be easy.

WHERE DO MOUNTAIN GORILLAS LIVE?

Mountain gorillas live in two small protected parks in Africa.

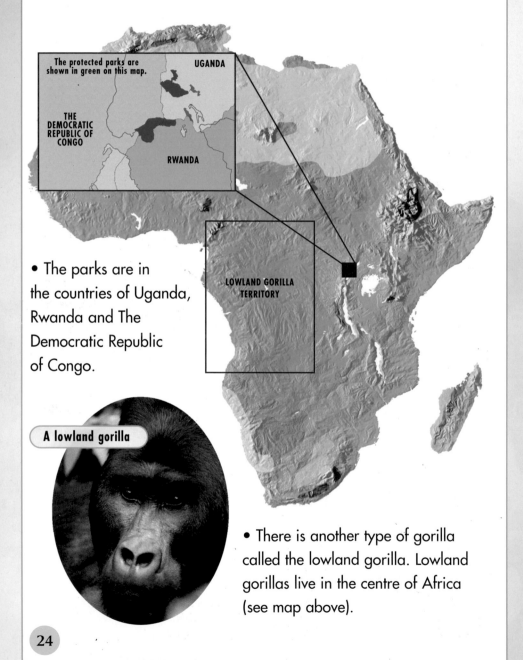

The protected parks are shown in green on this map.

UGANDA

THE DEMOCRATIC REPUBLIC OF CONGO

RWANDA

LOWLAND GORILLA TERRITORY

• The parks are in the countries of Uganda, Rwanda and The Democratic Republic of Congo.

A lowland gorilla

• There is another type of gorilla called the lowland gorilla. Lowland gorillas live in the centre of Africa (see map above).

MOUNTAIN GORILLA BODIES

*Gorillas are the largest of all the **apes**.*

• Gorillas walk on all fours, using their arms and legs. When they do this they are about 1 metre tall at their shoulders.

1 metre tall

Male
Weight: up to 200 kg

Female
Weight: up to 90 kg

2 metres tall

• A male gorilla can be 2 metres tall when he stands upright.

• A male gorilla's arm-span can be nearly 2.5 metres!

25

GORILLA FAMILIES

A gorilla family group

• Gorillas live in family groups. An adult male, called a silverback, leads the group.

• Young males usually leave the group when they are about 11 years old. By the time they are 15, they will usually have their own family group.

• Young female gorillas leave their group and join a new one when they are about 8 years old.

GORILLA FOOD

• Mountain gorillas are mainly **vegetarian**. They eat stinging nettles, bamboo, thistles, plant shoots and leaves, wild celery and blackberries.

A gorilla family in their nest of leaves.

• A diet of plants does not supply a lot of energy. Gorillas need to spend a lot of their time foraging and eating.

• Gorillas get almost all the water they need from eating plants.

GORILLA LIFE

• In the wild, gorillas live for 30 to 40 years.

• At night, gorillas sleep in nests on the ground. These are made from leaves and branches.

• Grooming is an important part of a gorilla's daily life. Gorillas clean each other's fur and remove insects.

A silverback grooms a female gorilla.

GORILLAS IN DANGER

All gorillas are in danger because of poaching and habitat loss.

HOW MANY GORILLAS?

Type of gorilla	Number left
Mountain gorilla	700
Eastern lowland gorilla	7,000
Western lowland gorilla	10,000

• Many zoos are helping to protect lowland gorillas. Mountain gorillas cannot be protected in this way because they die if they are kept in a zoo.

CONSERVATION

*Because of **conservation**, mountain gorilla numbers are now beginning to go up slowly. To save the gorilla, it is just as important to help the people who live near them as it is to help the gorillas themselves.*

• Conservationists show people how to farm rabbits and goats instead of catching meat with snares that can injure or kill gorillas.

• Local people can earn money from tourists who come to visit the mountain gorillas. This means they want to protect the gorillas and the forest.

Saving the mountain gorillas' habitat is not just a good thing for the gorillas. It helps all the other plants and animals that live in the forest as well.

HOW YOU CAN HELP THE GORILLAS

• Find out about mountain gorillas and other animals in danger. Do a project or a display at your school to tell other people about them.

• Join an organisation like the *International Gorilla Conservation Programme*. These groups need to raise money to pay for their conservation work. You could organise an event to help raise funds – try having a sale of all your unwanted clothes, old toys and books! See the websites below for lots of fundraising ideas.

• Be a good conservationist. Look after the place where you live. There are some good ideas to help you on the *Go Wild!* section of the *World Wildlife Fund* website.

Visit these websites for more information and to find out how you can help to 'Save the gorilla'.

The International Gorilla Conservation Programme: www.mountaingorillas.org

The Dian Fossey Gorilla Fund International: www.gorillafund.org

African Wildlife Foundation: www.awf.org

Fauna and Flora International: www.fauna-flora.org

World Wildlife Fund International: www.wwf.org.uk

GLOSSARY

apes A group of mammals that includes gorillas, chimpanzees, orang-utans and humans.

conservation Taking care of the natural world. Conservationists try to stop people hunting animals and they ask governments to pass laws to protect wild habitats.

dormant volcanoes Volcanoes that are not erupting now, but might erupt again in the future.

foraging Looking for food in the wild.

habitat The place that suits a particular animal or plant in the wild.

park rangers People whose job it is to look after the protected parks and the animals that live there.

protected parks Areas where hunting animals or cutting down trees is against the law.

rainforests Warm, wet places where lots of trees and plants grow. Many animals, such as antelopes, forest buffaloes and golden monkeys, live in the rainforests where the mountain gorillas live.

snares Traps made from a loop of wire. When an animal treads on the wire, it springs up and holds the animal's leg. The more the animal struggles, the tighter the wire becomes.

tourists People who are on holiday.

vegetarian An animal or person that does not eat meat.

INDEX